William Senior

Scotch Loch-Fishing

William Senior

Scotch Loch-Fishing

ISBN/EAN: 9783337412890

Printed in Europe, USA, Canada, Australia, Japan

Cover: Foto ©Lupo / pixelio.de

More available books at **www.hansebooks.com**

SCOTCH LOCH-FISHING

BY

"BLACK PALMER"

WILLIAM BLACKWOOD AND SONS
EDINBURGH AND LONDON
MDCCCLXXXII

All Rights reserved

Dedicated

TO THE

MEMBERS OF THE WESTERN ANGLING CLUB GLASGOW

IN REMEMBRANCE OF MANY HAPPY DAYS

SPENT IN THEIR COMPANY

PREFACE.

The Author of this very practical treatise on Scotch Loch-Fishing desires chiefly that it may be of use to all who read it. He does not pretend to have written anything new, but to have attempted to put what he has to say in as readable a form as possible. Everything in the way of the history and habits of fish has been studiously avoided, and technicalities have been used as sparingly as possible. The writing of this book has afforded him much pleasure in his leisure moments, and that pleasure would be much increased if he knew that the perusal of it would create any bond of sympathy between himself and the angling community in general. This edition is interleaved with blank sheets for the reader's notes. The Author need hardly say that any suggestions addressed to the care of the publishers, will meet with consideration in a future edition.

GLASGOW, *March* 1882.

CONTENTS.

CHAP.		PAGE
I.	INTRODUCTORY,	1
II.	EQUIPMENT,	5
III.	TACKLE AND ACCESSORIES,	7
IV.	FLIES AND CASTING-LINES,	13
V.	TROLLING-TACKLE AND LURES,	21
VI.	DUTIES OF BOATMAN,	27
VII.	ETIQUETTE OF LOCH-FISHING,	33
VIII.	CASTING AND STRIKING,	37
IX.	TROLLING,	42
X.	CAPTURE OF FISH,	48
XI.	AFTER A DAY'S FISHING,	60
XII.	REMINISCENCES,	65
XIII.	CONCLUSION,	80

SCOTCH LOCH-FISHING.

CHAPTER I.

INTRODUCTORY.

WE do not pretend to write or enlarge upon a new subject. Much has been said and written—and well said and written too—on the art of fishing; but loch-fishing *per se* has been rather looked upon as a second-rate performance, and to dispel this idea is one of the objects for which this present treatise has been written. Far be it from us to say anything against fishing, lawfully practised in any form; but many pent up in our large towns will bear us out when we say that, on the whole, a day's loch-fishing is the most convenient. One great matter is, that the loch-fisher is dependent on nothing but enough wind to "curl" the water,—and on a large loch it is very seldom that a dead calm prevails all day,—and can make his arrangements for a day, weeks beforehand; whereas the stream-fisher is dependent for a good take on the state of the water: and however pleasant and easy it may be for one living near the banks of a good trout stream or river, it is quite another matter to arrange for a day's river-fishing, if one is looking forward to a holiday at a date some weeks ahead. Providence may favour the expectant angler with a "good" day, and the water in order; but experience has taught most of us that the "good" days are in the minority, and that, as is the case with our rapid running streams,—such as many of our northern streams are,—the water is either too large or too small, unless, as previously remarked, you live near at hand, and can catch it at its best.

A common belief in regard to loch-fishing is, that the tyro and the experienced angler have nearly the same chance in fishing,—the one from the stern and the other from the bow of the same boat. Of all the absurd beliefs as to loch-fishing, this is one of the most absurd. Try it. Give the tyro either end of the boat he likes; give him a cast of any flies he may fancy, or even a cast similar to those which a "crack" may be using; and if he catches one for every three the other has, he may consider himself very lucky. Of course there are lochs where the fish are not abundant, and a beginner may come across as many as an older fisher; but we speak of lochs where there are fish to be caught, and where each has a fair chance.

Again, it is said that the boatman has as much to do with catching trout in a loch as the angler. Well, we don't deny that. In an untried loch it is necessary to have the guidance of a good boatman; but the same argument holds good as to stream-fishing. There are "pools and pools," and the experienced loch-fisher can "spot" a bay or promontory, where trout are likely to be lying, with as much certainty as his brother angler can calculate on the lie of fish in a stream. Then there are objections to loch-fishing on the score of expense. These we are not prepared to refute; for there is no doubt whatever that loch-fishing means money. But what has made it so? The same reason that makes all other things of more or less value—the common law of supply and demand. Time was, and that not so long ago, when a boatman who used to get 3s., or at most 4s. a-day, now gets his 5s. or 6s., and even at the latter figure does not think himself too well paid. In the extreme north, however, it is still possible to get a good man for 3s. a-day; and we know of nothing more enjoyable than a fortnight's loch-fishing amidst magnificent scenery in some of our northern counties. The expense of getting there will always be a serious matter; but once there, the fishing in itself is not dear. The boat is usually got for nothing; the right of fishing, so far at least as trout are concerned, is free; and the man's wage and lunch are decidedly cheap. But for a single day on some of our nearer lochs,—such as Loch Leven, Loch Ard, or Loch Lomond,—the expenses *are* heavy, and the angler must always be the best judge as to the likelihood of the "game being worth the candle."

CHAPTER II.

EQUIPMENT.

THIS will be a short chapter, as tastes differ so very much, that many things we might say would most probably be disregarded. But as to some matters, there can only be one opinion. Do not fish in *light-coloured* clothes; and, should the weather be wet, do not wear a white macintosh coat. We believe that the eyesight of a fish is the keenest sense which it possesses; and, more especially should the day be clear and fine, there is no doubt that an unusual white object within range of its vision will make a fish, which might otherwise have taken the fly, turn tail and flee. A good deal of what we hear spoken of as fish "rising short," proceeds from this cause. No doubt they rise short sometimes on seeing the angler himself, but he is much less likely to attract notice if clad in dark-hued clothing. We know of nothing better for a fishing rig-out than a suit made from dark Harris tweed—it will almost last a lifetime, and is a warm and comfortable wear. Thus you will need a dark macintosh and leggings; and a common sou'-wester is, when needed, a very useful head-gear. A pair of cloth-lined india-rubber gloves will be found desirable in early spring, when it is quite possible that the temperature may be low enough for snow. A pair of stout lacing boots, made with uppers reaching well up the leg, will be found best, as they protect the feet from getting damp when going into or leaving a boat, even though one should need to step into the water; and if your waterproof coat is long, as it should be, the necessity of wearing leggings on a wet day is obviated. Lastly, *by all means keep the body warm*, and remember that the more careful you are of yourself, even at the risk of being thought "old wifish," you will, humanly speaking, be enabled to enjoy the sport to a greater age than you might otherwise do.

CHAPTER III.

TACKLE AND ACCESSORIES.

AS this is likely to be one of the most important chapters in the book, the reader must forgive us if we are particular—even to a fault—in describing some of the necessaries towards the full enjoyment of the pleasures of loch-fishing. So much depends on our being comfortable in our enjoyments, that we have, perhaps, erred on the side of luxuriance; but to those anglers who think so, there is nothing easier than their leaving out what they think superfluous.

Creel, or Fishing-Bag.—The creel for loch-fishing should be of the largest size made, so as to serve for all kinds of fish; and as the angler is always in a boat, the difference of room occupied is of very little moment. Besides, it accommodates his tackle and lunch, and even waterproofs, though the latter are better to be strapped on outside. These creels are neatest when made in French basket-work; and even the lightest of them, with ordinary care, will last many years, more especially if the edges and bottom are leather-bound. Almost any tackle-shop will supply them plain, or bound with leather, as desired. Brass hinges and hasp will also be found great improvements. The fishing-bag is of somewhat recent development, and is very convenient; but the objection to it is that, unless the waterproof cloth with which it is lined be carefully washed after each day's fishing, a nasty smell is apt to be contracted and retained. Though we use the bag often ourselves, we incline for many reasons to the old-fashioned creel. Many loch-fishers carry along with them a square basket about 16 in. × 8 broad × 10 deep, which they use for carrying their tackle and lunch, thus leaving the creel or fishing-bag free for fish alone. This is a capital plan, the only objection being that it makes another article to carry. As to its usefulness there can be no doubt, as nothing is more undesirable than having tackle and fish in one basket or bag, even though you should have something between. Some anglers go the length of a luncheon-basket, but this savours so much of the picnic that we don't approve of it.

Landing-Net and Gaff.—These may be got at any tackle shop, the only care to be exercised being in the selection of a good long handle, and in seeing that the net be made of twine which resists the catching of hooks, and that it be of a size capable of landing a large fish, as the gaff leaves an ugly mark, and should only be used when actually necessary. The screw of the net-hoop and of the gaff will suit the same handle.

Fishing-Rods.—For loch-fishing, it is desirable to use a rod not less than 14 feet in length, if fishing for ordinary yellow trout; but if for sea trout, and the chance of "a fish" *par excellence*, then the rod should be a couple of feet longer. The angler will find that it is better to have both rods with him—the spare one being handy in case of calamity—as the extra trouble of carrying is very slight: rods and landing-net handle can be easily tied up together with small leather straps. Do not have a rod that bends too freely—rather err on the other side; because in loch-fishing you have generally wind enough to carry your flies out, and if you do get a 3 or 4 pounder, the advantage of a fairly stiff rod is apparent. We prefer rods in three pieces—no hollow-butts—and made of greenheart throughout. The first cost is more than for rods whose various parts are made of different woods, but the greenheart is the cheapest rod in the end. With the minimum of care, a greenheart never gets out of order; and a good rod of this description will be as straight at the end of a season as at the beginning. Avoid all fancy rods, and do not be beguiled into buying them.

Reels and Lines.—Always carry a couple of reels with you, the smaller with 60 yards of fine line, and the larger with not less than 100 yards of grilse line. Silk-and-hair lines are not very expensive, and with a little care will last a long time. They will be found the most satisfactory for all kinds of fly-work. The reels which we consider best are made of bronzed metal and vulcanite: they are light, and stand a lot of wear. When buying your rods, get the reels fitted to them, and see that the fit is sufficiently tight, as nothing is more annoying than to find the verrules loosening their hold of the reel, and that, perhaps, at a most critical moment. Should the reels referred to not be heavy enough to balance the rods properly—and this is a matter of great importance—it may be as well to take reels made entirely of bronzed metal.

Fly-Book.—We are not much in favour of fly-books. They are a great temptation to keeping a large stock of flies; and in the following chapter we will show that the fewer flies one possesses the better. A serious objection to a fly-book is, that the flies get crushed in it, and we consider a box a better receptacle; but if the angler will have a fly-book, one of moderate size—rather to the big size if anything—made of pig-skin leather, and well provided with pouches for holding casting-lines, as well as the usual receptacles for flies, will be found best. These books are to be had in great variety at any wholesale tackle warehouse; and taste goes a long way in non-essentials.

Beyond the articles mentioned, the angler should always have at hand the following:—

> Spring balance, weighing up to 20 lb.
> Small screw-driver.
> Small gimlet.
> Small bottle clockmaker's oil.
> Bottle varnish.
> Carriage-lamp, and candles to fit, for travelling.
> Two packs playing-cards.
> Good-sized flask.
> Flat glass or horn drinking-cup.
> Pocket-scissors. The kind that shut up will be found very useful.
> Corkscrew.
> Hank of medium gut for emergencies.
> Fine silk thread and resin.
> Some common thin twine for tying joints of rod together.
> Also articles named in Chapter V., p. 21, under "Trolling-Tackle and Lures."

Many of these things may be considered quite *de trop;* but the longer one fishes, the more one finds out that the little luxuries give a vast amount of enjoyment for the small amount of foresight required to have them at hand when wanted.

CHAPTER IV.

FLIES AND CASTING-LINES.

FLIES.—Here we shall no doubt come into conflict with many opinions, and most probably meet with the most criticism. However, as all we have written, and mean to write, is the result of actual experience, we may be pardoned for being somewhat dogmatic on the subject in hand. In the first place, don't keep a large stock of flies. If going for a day's fishing, buy as many as you think you'll need, and *no more*. Buy them of different sizes; and if you get a few each time you go for an outing, you will be astonished how soon a spare stock accumulates. Ascertain carefully beforehand the *size* suitable to the loch—the *kinds* are not of so much importance—and once you have made up a cast, in which operation there is no harm in taking your boatman's suggestions, *do not change*, unless it be to put on a smaller or larger size according to the wind, or unless it is conclusively proved that other flies are raising trout when yours cannot. Of course, if you are going for a fortnight's fishing, you will require to lay in a fair stock; but even then get as few as you think you can possibly do with. Do not run any risk of running short, and do not place yourself in the position of needing to use old casts: that is poor economy in the long-run. The following is, we think, a fair list for a fortnight's sport in an out-of-the-way place :—

Half-dozen harelugs.
- „ red and teal.
- „ orange and mallard.
- „ green and woodcock.
- „ black spiders with red tips, commonly called "Zulus."
- „ red spiders, hackle taken well down the hook.
- „ March Browns, which, though supposed to come out in March, are really capital flies at any time.
- „ yellow body with cinnamon wings and golden-pheasant tip.
- „ dark harelug body, mallard wing and red tip. This is a splendid spring fly.

These we would get dressed on Loch Leven size—any fly-dresser knows what that means; but perhaps the better way would be to get a quarter dozen of each dressed on that size, and a quarter dozen of each on a hook two sizes larger. The patterns in a tackle-maker's book are endless, but for the most part are modifications or combinations of the flies we have named; and the angler will soon discover for himself that flies and old half-used casts, and often casts made up in the humour of the moment, and never used at all, accumulate upon him so rapidly that he is glad to find some enthusiastic boatman to bestow them upon. It is needless to add, that a gift of this kind is usually very much appreciated by the recipient. Tinsel is a very useful adjunct to a fly, and should always be employed in those used in loch-fishing. If variety is wanted in colouring, the least tip of Berlin or pig's wool of the desired shade will be found very effective. Get your flies dressed on Limerick-bend hooks, as the iron, should it chance not to be the best tempered in the world, is not so liable to snap as the round bend. The wings of the fly should be dressed so as to be distinctly apart both in

the water and out of it, thus—

It gives the fly a much more life-like appearance, and makes it swim better in the water. When you give orders for flies, see that they are dressed up to your instructions, as it is quite certain you will fish with much more confidence when you have faith in what you are using. Do not have them dressed on too fine gut, as they are apt to get twisted round the casting-line (usually called "riding the line"), and put you to the trouble of straightening them out every few minutes. These remarks may seem trifling; but trifles are very irritating in most pursuits, and the gentle art is no exception. Flies suitable for salmon and sea-trout fishing on almost any loch will be supplied at any shop in the trade on asking for Loch Lomond patterns. These patterns are well-known, and are without exception as fine flies as one could wish for. They are usually made very full in the body, and dressed with heron's hackle. The varieties are red and teal, green and teal, orange and mallard, or turkey, and a few variations of these,—sometimes a yellow tip to the red and green bodies, or a red tip to the yellow; but a cast composed of red or green and teal with orange and mallard is unsurpassable. For this class of fishing, the flies should be dressed with loops, and the bob should be fixed to the casting-line by means of a small strand of gut. Two flies on a cast are quite sufficient when big fish are expected. We can hardly advise the angler to try fly-dressing on his own account. It is hardly worth his while, as flies are to be had very reasonably from any respectable tackle-maker; and they are much better dressed in ninety-nine cases out of a hundred than any amateur performance.

Casting-Lines.—Provide yourself with half a dozen each, of different thickness—that is, fine, medium, and stout, the latter for salmon and sea-trout fishing. That quantity should suffice for a fortnight's outing, even making allowance for breakage, and leave you some over for another time: but in this matter it is better to run no risk of being short. The gut should be stained a light tea colour, or the faintest blue: it can be bought so. There is no occasion for them being more than three yards long, as we cannot advocate fishing with more than three flies at a time. If three flies are properly placed on a line, and the line be properly handled in the casting, they will cover as much water as any number of flies. Besides, there is far less chance of a "fankle," to use a most expressive Scotch word, than when four or more flies are used. In this, however, *chacun à son goût,*— we are only giving an opinion after trying both ways.

In making up a cast of flies, *have no loops* of any kind, excepting the one by which the cast is attached to your silk-and-hair line. The water-knot is so simple and neat, that it is the best for the purpose of fixing on the tail-fly, which, by the way, should be the heaviest of those you are about to use, if there is any difference between them. In case our readers don't know the water-knot, we give an illustration which explains itself—

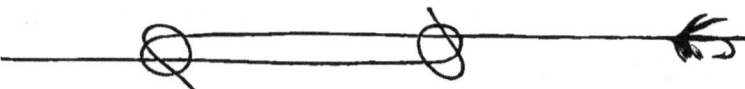

The loops are pulled tight, and then the fly and the line are drawn in opposite directions, the result being that the knots formed by the loops meet and make a firm, and at the same time an almost imperceptible, joining. You then clip off any ends that may remain. So much for the tail fly. The putting on of the other two is simplicity itself. You take the strand of gut on which the next fly you purpose affixing is dressed, and laying it along the main line, *taking care to have the hook lying in the reverse direction from the tail fly*, you tie it into the line a yard from the fly already attached. In tying it in, leave the hook hanging about two and a half inches from the line. The third, or "bob" fly, is attached in like manner, and thus your cast of flies is completed. The only objection to this method of making up a cast is, that once the middle and bob flies are tied in, they cannot be used again. This is quite true; but the keen angler will submit to the little extra expense on this score for the gratification which the sight of a really neat cast will afford him. The system of suspending hooks by loops, especially when using fine tackle, is almost entirely exploded. We should have said that previous to use, all gut should be soaked, and the longer the better. It is a good plan to let it soak over night, and make up your cast in the morning. When gut is thus thoroughly wet, it is wonderful how easy it is of manipulation. On the other hand, dry gut is very brittle, and will break on the slightest provocation. Fix the cast and silk-and-hair line together, having previously made a single knot on the end of the latter, as illustrated below—

It is prudent to have a second cast ready in case of breakage, as nothing is more annoying than losing time making up one in the boat, and that most probably when the trout are rising. Experience is a great teacher; and it is wonderful how soon the angler learns the value of every moment, and seeks beforehand, so far as human foresight can go, to provide for all contingencies.

CHAPTER V.

TROLLING-TACKLE AND LURES.

Do not troll at all if you can get fishing with the fly; and under no circumstances troll for trout in the very early part of the season, when they are more or less in a "kelty" state, and take an artificial or other minnow very keenly. True, you may catch fish, but it is a most unsportsmanlike proceeding to take fish not in fair condition; and if you, sir, who read this book, are not a sportsman, you had better stop here, for it was compiled by a sportsman for sportsmen. There are some miserable "pot-hunters" who want to kill anything that swims—be it clean or unclean; but with them we have nothing whatever to do. But fair trolling is quite legitimate, and in many cases it is absolutely imperative to troll if a basket is to be made at all. Some days the fly is of no use—either owing to a calm, or to a bright sky; and a well-managed trolling-line or two is then the only resort, unless one stops fishing altogether. And if big trout and *ferox* are wanted, nothing succeeds — indeed nothing *will* succeed, except very occasionally —but trolling, either with artificial or natural bait. So to be complete you must have the requisite tackle, and we will tell you what is necessary, both for small and large fish, in as few words as we possibly can. A ROD specially adapted for trolling is almost a necessity, as it is a great strain upon an ordinary fly-rod to have the weight of 30 or 40 yards of line upon it: even a good rod is apt to get an ugly bend from such treatment. The rod for trolling need not be long—12 to 14 feet is quite sufficient—but it must be stiff; and we consider that the rings through which the line is led ought to be large and fixed—that is, standing out permanently from the wood, called by the trade upright rings. A spare top will be supplied along with it. The REEL should be of the largest description, and may be got as strong as possible, lightness being no recommendation to one used exclusively for trolling. The LINE ought to be at least 100 yards long—120 for choice—and this suffices for any kind of fish. The material best adapted for trolling is oiled silk-and-hair. There is a kind of line, made in America, we believe, which is admirably adapted for the purpose. It is strong as wire-rope, and does not "kink" under any circumstances—which latter is a consideration, as sometimes a paltry trout may come on, and you have only to haul him in hand-over-hand without running the risk of your line getting into a mess. This saves the trouble and waste of time in reeling up many yards of line every time a "smout" comes on. The line to which we refer is somewhat expensive, but will be found to be cheap in the long-run. An ordinary silk-and-hair line does well enough, but is apt to twist sadly if the minnow is not spinning properly, besides the trouble it entails after a day's fishing of laying out two or three score yards for drying. The troller will require to provide himself with MINNOW TRACES. These do not require to be more than two yards in length, but in ordering them take care that the swivels are sufficiently large to insure the minnow — natural or artificial — spinning nicely. The angler can easily procure swivels and make traces for himself; but he will find in this, as in most things connected with fishing, that he cannot compete with the tackle-maker, so we advise him to get them made up at a good warehouse. Retail tackle-makers charge long prices, but in most large towns there are warehouses which are specially suited for a customer trade, thus saving the user a long intermediate profit. This is as it should be. The thickness of the gut used for trolling should of course be regulated, as in fly-fishing, by the size of fish you expect to catch, and a few traces made of gimp for pike and *ferox* should always be in the troller's stock. By the way, and in case we forget to mention it afterwards, always be provided with some split swanshot, to be used in case of a very clear day, when it is desirable to sink line and minnow below the surface. Also be provided with tackle—some mounted on gut and some on gimp—for spinning natural minnow; and we know of none better or more deadly for this purpose than that of which an illustration is given on next page. It is very simple, and seldom misses anything.

The large hook is put in at the mouth of the minnow, and the point brought out at a little

above the tail—thus giving the minnow the proper curve for spinning. One of the smaller hooks is put through *both* lips of the lure, to close the mouth and to keep the bait in proper position, while the other is left to spin. Some advocate the use of par-tail as a spinning bait; but as it is not right to kill par at all, we omit any directions for its use. We have drifted into the subject of LURES almost unconsciously. If you wish to use natural minnows, see that they are neither too large nor too small—about two inches long is a good size — and that the belly is silvery. It is better to instruct your boatman to have a supply ready against your arrival at the loch, as sometimes it is as difficult to catch minnows as to catch any other fish. However, we believe that the want of them is so well supplied by the phantom minnow, that little or no harm is done though they are not to be had. And when the handling and bother of using live bait is taken into consideration, we think that most folks will prefer the artificial lure. The phantom we consider the very best of all the imitations; and the troller should have them in different colours and sizes, from Nos. 1 to 7. The hooks attached to the larger sizes should be mounted on gimp, as in trolling for large fish—and especially for *salmo ferox*—no risk should be run of the mountings giving way. Tin boxes, divided into compartments, for holding the minnows, are very convenient, and are to be had at most tackle shops. A spoon-bait is also a splendid deception, and should not be awanting. A tackle-maker's catalogue will tell the reader of many other "spinners;" but if he cannot catch fish of all kinds with either a natural or phantom minnow or a spoon, it is not the fault of the lure; and he may try anything else he fancies, and come no better speed.

CHAPTER VI.

DUTIES OF BOATMAN.

VERY little requires to be said in this chapter regarding boatmen, as when the angler gets into the habit of frequenting certain lochs, he soon finds out for himself the steady reliable men in the neighbourhood, and can generally engage one of them beforehand by writing to the hotel at which he means to put up. But in going to a new fishing-ground, he is better to leave himself in the hands of the landlord of the hotel, and if not satisfied with his first day's experience of the man who accompanied him, let him change. A good boatman is a treasure; and though we are decidedly against the system of "tipping" indiscriminately, we say, when you get a good man, pay him liberally. We know of some men with whom it is a pleasure to be out all day, and whose company, in its own way, is most enjoyable. Keen sportsmen these are, and the capture or loss of a fish is a source of true pleasure or pain. Other men one comes across seem but to row the boat, and nothing more; and an unproductive day in such company is something to be looked back upon with horror. The leading qualification of a boatman of the right sort is a strong sympathy with the angler, which enables him almost instinctively to help the angler to cover every inch of likely water with his flies, and makes him experience the sensation of expecting a rise every cast; in other words, he almost puts the fly into the fish's mouth. With such a man, instructions regarding the management of the boat are superfluous; but as it often happens that you do not get a first-rate hand, you have to take matters into your own hands to some extent; and we shall give you a few hints as to what is best to be done under such circumstances. It is hardly to be supposed that your man is in ignorance of the best ground, either from experience or hearsay, and it is only after you get there that our instructions can possibly come into operation. If you are obliged to take a perfect greenhorn, we know of no other course than to order him to keep in the wake of some other boat, but that at such a distance as not to be offensive. (See next chapter on the "Etiquette of Loch-fishing.") But let us assume that you get on to ground where fish are: the first point is to see that everything is in order, all unnecessary articles put out of the way, and the landing-net and gaff conveniently at hand. We ought to have said that a large stone in the bow is useful, not only to balance the boat and make her drift better, but also as a weight to which a rope may be attached, and thus let over the side to the depth of a few feet, to prevent her drifting too rapidly should there happen to be a heavy breeze on. The next thing is to get the boat properly broadside to the wind, so that you may have next to no trouble in casting. Should a fish be hooked, see that the man keeps working the boat in such a manner that the fish cannot possibly get underneath: a single stroke of the oar in the proper direction is generally all that is necessary. You must also judge from the size of the fish, and the length and strength of your tackle, whether it is expedient that the man should follow the fish if he makes a very long run. If your line happens to be short—which it will not be, if you have followed the instructions given in Chapter III.—you need not be surprised if you find nothing left but your rod and reel, your line, and mayhap a "half-croon flee" flying about the loch in charge of a fish. The management of the landing-net or gaff is another serious matter. If the fish be small, tell the man to have the net ready, and "run it in;" but if it is a good-sized fish, you must tell him not to put the net near till he gets the word from you. Many a time we have suppressed an exclamation—the reverse of a blessing—when we have seen the hoop of the landing-net strike the fish, and were in suspense for a second or two as to whether he was on or off. If the gaff is necessary, it is almost as well to let your man hold the rod after you have tired the fish thoroughly, and gaff him yourself. But if you think it unadvisable to part with the rod, send the man to the other end of the boat from yourself, and then lead the fish near him, so that he may have a fair chance. He must put the gaff *over* the fish till the point is in a line with its broadside, and then with a sudden *jerk* sink the steel into, or even through, the animal, and lift him over the gunwale with all possible speed. A sharp blow or two on the snout will deprive the fish of life. Always kill your fish, —big or small,—as nothing ought to be more repulsive to a true sportsman than to see or hear any animal he has captured dying by inches.

It is perhaps needless to say that in the matter of lunch and drink, due consideration should always be paid to your boatman's wants; indeed if he has had a hard time of it rowing against a stiff breeze, nothing is lost by landing at mid-day and letting him enjoy half an hour's rest and a smoke after he has refreshed his inner man. Sometimes — such as in a club competition — such luxuries must be denied; but even then he can put you on to a square drift, and enjoy his lunch and smoke while you are fishing; and you, on the other hand, can take yours when he is changing ground. These remarks may seem trifling; but we only give you our experience, when we say that on some lochs where good boatmen are not plentiful, the angler who has shown most kindness and consideration on past occasions is never much put about for want of a man, even in a busy season. And we have known, when every regular boatman was engaged, that there was generally a boatman's "friend" in the neighbourhood who was pressed into our service, and that often at a few minutes' notice.

CHAPTER VII.

ETIQUETTE OF LOCH-FISHING.

POLITENESS is politeness all the world over, and in loch-fishing it is particularly to be practised. The gentle art is peculiarly adapted for gentlemen,— using the word in its truest sense,—and the true angler will never be mistaken for anything else. In the Club to which we have the honour to belong, there are certain rules which would commend themselves naturally to any one of us; but in order that these may be clear and well defined, they are circulated annually, and are in themselves so admirable that we cannot do better than quote them :—

"1. No boat shall be entitled to take position in front of any other boat which shall have already begun drifting, at a less distance than three hundred yards.

"2. Any competitor intending to drift a bay already in possession of another competitor shall be obliged to take position behind, or on the outside of and in a line with the latter, but at such a distance as not to interfere with the boat first in possession of the drift.

"3. In cases where boats are changing water, it shall not be admissible for any boat so doing to go between the shore and any other boat drifting close thereon."

These rules, as may be inferred, refer to club competitions in particular, but they are made the standard upon all occasions where there is any chance of their becoming applicable. So much indeed have we got into the way of regarding these rules,—strict as they are,—that we observe them even when meeting with strangers on any loch in any part of the kingdom. And pay special attention, if you happen to be trolling in the neighbourhood, never to interfere with the drift on which a fly-fisher is engaged. Nothing is more unbecoming, as it disturbs the water which is his by right, if he has begun to drift; and it is an unwritten rule that the fly-fisher should generally be allowed the first of the ground, as his style of fishing does not make the same commotion as a trolling boat and tackle do. Very few of us but have experienced the annoyance of a minnow-boat crossing our drift when we were fly-fishing; and though we had no redress, and could make no remarks without lowering ourselves to the level of our offenders, we have, like the nigger's parrot, "thought a mighty lot." Do not hesitate to put yourself out of your way to help a neighbour in distress. He may have hooked a large fish and be unprovided with a gaff: if you have one let him have it instantly, taking his directions from which side you are to approach him; and never let the loss of a few minutes, more or less, deter you from following the golden rule of doing to him as you would expect him to do to you were you similarly placed. And, as it sometimes happens where boats are scarce and anglers many, when you are in the same boat with a stranger, see that you confine yourself strictly to your own share of the water, not making casts which endanger "fankling" for the mere sake of covering a little more water with your flies. Should you have a fly that is taking better than any other of your own or his, offer him one; and in general try to make the day's fishing one as much for the cultivation of goodwill and the promotion of good-fellowship as for the mere sake of making a basket. A churlish angler is an unnatural phenomenon, and, thank Providence! they seldom turn up. A man who can look upon the beautiful scenery amid which he takes his pleasure,—and there is none finer in the world than our Scottish lochs and their surroundings,—and not feel grateful to the Giver of all good, and at peace with all mankind, ought to burn his rod, singe his flies, and only associate with men like himself.

If the introduction of this chapter into our book will have the effect of creating a better understanding on the etiquette of loch-fishing between brothers of the angle, the object for which it was written will have been accomplished—and, let us hope, a large amount of goodwill thereby promoted.

CHAPTER VIII.

CASTING AND STRIKING.

WE shall treat this subject under two aspects: first, if you have the whole boat to yourself; and second, if it is being shared by some one else.

If you have a boat to yourself, stand as near the centre of it as you possibly can without interfering with the boatman in rowing, and cover every inch of the water in front of you and as far to the sides as the wind will permit. Always be careful how you cast—that is, every time you throw your flies see that they land lightly on the water, as no one can expect to raise fish if any splash is made by either line or flies. Fine casting is not quite so essential, of course, when a fair breeze is blowing; but if the wind be light, then the difference between a well-thrown fly and the reverse is very apparent. After you have made a satisfactory cast, draw the line slowly to you by raising the point of the rod, taking care to keep the line as taut as possible. Also see that your bob-fly is tripping on the surface, as we consider that a well-managed "bob" is the most lifelike of the whole lot. Do not fish with too long a line, unless, indeed, on an exceptional occasion, when you wish to reach the lie of a feeding fish. It is difficult to define a long line, but a good general rule is that it should never be longer than when you have the consciousness that, if a fish should rise, you have him at a fair and instantaneous striking distance. Remember that the time the flies first touch the water after each cast is the most deadly; therefore, cast often.

If you have only the share of a boat, the rule is that one man takes the stern up till lunch, and the other after it. For ourselves, we have a preference for the bow, and we generally find that most anglers prefer the luxury of the stern; so when both parties are pleased, there is no occasion for changing at all. The most important thing to bear in mind when you have a companion is, as we said in last chapter, to confine yourself to your own water. If the left-hand cast is the one proper to your end of the boat, cast as much to your right hand as you can without infringing on your neighbour's share of the water: all the water to your left hand is of course yours. The same remarks apply *vice versa*. *Never stop casting* so long as you are on fishable ground, for you know not the moment a good fish may rise. Certain it is that unless you keep your flies constantly going, you cannot expect to have the same basket as the angler who does. Keep your eyes on your flies in a general way, and do not let your attention be distracted so long as they are in the water. Every angler has experienced the annoyance of missing fish when looking elsewhere for a single moment—either at another boat, or at a fish " rising to itself," or at the sky, or at something else. When the eyes were turned to the point from which they should not have been diverted, they were just in time to see the water swirl, and the hand gave a futile strike at what had disappeared a second before. Perhaps we should have said at the beginning of this chapter to place implicit faith in the flies with which you are fishing. Nothing is more ridiculous than whipping the water with a cast, of the suitableness of which you have any doubt; and to guard against any such chance, study carefully the state of the weather and the wind. If very clear, use sombre flies; but if a dark day, use brighter flies. You will of course regulate the size according to the breeze, but as a rule, err on the side of small flies. When you raise a fish, *strike at once.* It is quite possible that by this method you may once in a while strike the least bit too soon, but it is a safe plan to go by. There is always a particle of a moment spent in the tightening of the line; and by the time the angler sees a fish at his flies, he may safely conclude that it has already seized or missed them, and the sooner he ascertains the true state of matters by striking instantaneously, the better. If the fish has not been touched by line or hook, cast gently over him again: the chances are that there will be another rise, and, if the fish has been feeding, every likelihood that the second or even a third time may be lucky. In striking small fish, the least tightening of the line is sufficient; but with large fish, when your tackle and hooks are strong, strike *firmly home* to send the steel well in, right over the barb. Tackle that will not stand this had better be given away or destroyed,—the latter for choice.

CHAPTER IX.

TROLLING.

OUR readers will have guessed, from what has preceded this chapter, that we don't believe in trolling if it can be avoided; but still there are times and occasions on which it must be practised, and we plead guilty to having gone in for it oftener than once, when we saw that fly-fishing was useless. On the other hand, however, we have set out with a firm determination to do a fair day's trolling,—and nothing but trolling,—but somehow or another it has generally ended in fly-fishing when we could, and trolling as a *dernier ressort* when we could not. This, we doubt not, has been the experience of many of our angling friends to whom the mere killing of fish is a secondary consideration compared with the enjoyment of real sport. But when trolling is the order of the day, either from choice or necessity, then this is the way to go about it. We assume, of course, that the angler is equipped with tackle and lines specified in Chapter V., and that he has a supply also of live minnows with him. The elaborate tin-cans for holding minnows are quite unnecessary so far as loch-fishing is concerned; any ordinary vessel will do well enough for a day, provided the water is changed now and again. In trolling, two rods will be found ample. They should be placed at right angles to the boat,—the "thowl-pin," or, if there is not one near enough the stern, anything (a cheap gimlet answers admirably) fixed into the gunwale, being sufficient to keep the rod in position,—so that the spinners, of whatever kind they may be, will be as far apart from each other as possible. Take care that the butts of the rods are well at the bottom of the boat, as we have seen a rod not sufficiently fixed go overboard before now. A main point in trolling is to have plenty of line out. There should never be less than thirty yards out from one rod, and not less that forty from the other. By this means, should a fish not see the first lure, he may see the second. If trolling with natural minnow, which is much more apt to get out of order than artificial ones, see that the bait is intact and spinning properly. This involves the trouble of hauling it in for examination now and then; but it is better to be at that trouble than be fishing with, mayhap, a mangled lure, or one that has got out of spinning order, and more likely to act as a repellent than an attraction to any fish in the neighbourhood. In trolling any likely ground, the proper way is to tell your man to zigzag it, not pulling the boat in a straight line, but going over the ground diagonally, and thus covering as much of it as it is possible to do with a couple score yards of line behind. The turning of the boat necessitates a considerable circle being taken to keep the lures spinning, and so that the lines do not get mixed up; and your man, after making the turn, should row in a slightly slanting direction towards the point from which he originally started, thus—

Trolling. 45

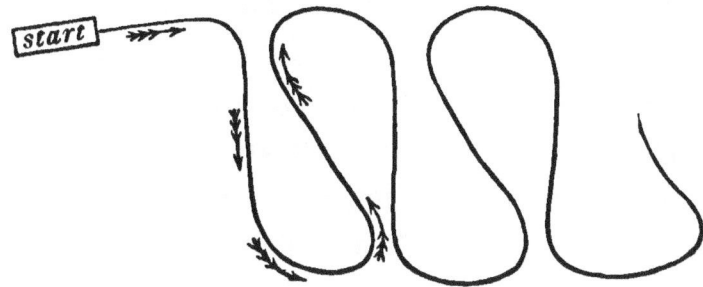

and so on, till the chances of raising a fish on that beat are exhausted.

Should a small fish come on, haul it in hand-over-hand; and the man must not stop rowing, as the other minnow is out, and must be kept spinning. If, however, a fish that needs playing comes to you, you must seize the rod to which he has come, and the boatman must take the other, and wind in as fast as possible. You should not commence winding in till the other line is wound up so far as to preclude the chance of the fish mixing up both lines together. Barring the risk one runs of a serious mess, it is not a bad plan to troll from a reel a cast of larger-sized flies than would be used in ordinary fly-fishing. This line follows, of course, in a *straight* track behind the boat, and the minnows being considerably to right and left of it, there is no danger of their getting mixed so long as the boat is moving; but the risk is apparent should a fish come to either of the three lines, and great activity is then necessary on the part of yourself and boatman to keep things right. You must keep the fish at as considerable a distance from the other lines as you can, and trust a good deal to the chances of war for the ultimate safety of all. Some days, even when casting was unproductive, we have been fortunate in securing fish by trolling our flies in the manner described. Indeed, unless the day or the season is decidedly in favour of trolling minnows, we prefer, if only trolling two lines, to troll from one of them with the minnow, and from the other with the fly. This must always be decided, however, by the judgment of the angler, and by his surroundings for the time being. One thing in favour of trolling with the minnow is, that the best size of fish are caught by that means. This is not invariably the case, but it is the rule. And in concluding this chapter, we must not omit to acknowledge that we are glad to know that when we are not so young as we once were, and when the wielding of a rod all day long shall have come to be a serious matter, we shall still have the pleasure of roaming about our lovely lochs—Highland or Lowland—and have the excitement of landing fish, coupled with our enjoyment of fresh air and grand scenery. For this reason, if for no other, cultivate as often as you can, without entrenching on the nobler pastime of fly-fishing, the art of trolling—for we must confess that there is an art in this as in everything else; and should my reader be sceptical on the point, he has only to try conclusions, when he gets the chance, with some old troller, and he will be convinced before supper-time.

CHAPTER X.

CAPTURE OF FISH.

SCOTCH loch-fishing, as usually practised, only embraces the capture of the *salmo* species—that is, the *salmo fario*, or common yellow trout; the *salmo trutta*, or sea-trout; and *salmo salar*, the "fish," as most boatmen call it, and the noblest game of the finny creation. Besides these there is, of course, the *salmo ferox;* but it is comparatively scarce, and only worth trolling for in some particular lochs, where they are known to be more easily come across than in others. And sometimes when worthier game is not to be had, we have a spin for pike, but Mr Jack is as difficult to catch at times as his more aristocratic comrades. In most Scotch lochs where any supervision is exercised at the instance of our local clubs, the extermination of pike is most vigorously carried on by means of fixed and splash nets. This, as regards our large lochs, where there is room for all, we have no hesitation in saying is a mistake, as it shuts up one means of enjoying a day's fishing when nothing else in the way of fish is to be had; and it must be borne in mind that there are some older anglers, to whom a whole day's fly-fishing is a labour, who never object, when trolling, to come across a pike: and no wonder, for a pike of 10 lb. and upwards gives some fair play, though by no means to compare with what a fish of the *salmo* tribe of that weight would give. Then we have perch in abundance, and splendid eels; but as these need a float and bait to catch them, we dismiss them as quite *infra dig*. True a perch will come at a minnow, and we have sometimes seen them take a fly; but they are generally voted a nuisance, and expelled the boat.

As regards the capture of fish, we shall proceed to deal with each in order; and at the outset we remark, that when you have hooked a fish, it is a safe general rule to waste no unnecessary time in bringing him to the landing-net or gaff, and thence into the boat. When playing a fish, never allow the line to get slack, unless, indeed, when he leaps into the air,—then you must give him rope; but so soon as he gets into his native element, feel his mouth instantly. Always play your fish to *windward* of the boat if there is some one sharing it with you, as this allows him to go on casting to leeward. Of course, if you have the whole boat to yourself, play your fish in any way that it will be most expeditiously brought to basket. The angler ought to be well assured of the strength of his tackle, and when he has confidence in that, he will soon learn to judge of the proper strain to which it may be subjected. In the case of COMMON YELLOW TROUT, averaging, as most loch trout do, about three to the pound, there is no occasion to put off time with any one of them; but in some lochs, such as Loch Leven, where the average is fairly one pound, and where two and three pounders are by no means uncommon, some care and a little play are absolutely necessary. But do not, even in such a case, give him too much of his own way. We can assure our readers that a three-pound Loch

Leven trout, in good condition, on fine gut and small irons, gives as nice a piece of play, and exercise to the eye, hand, and judgment, as could well be desired.

The SEA-TROUT is, for his size, the gamest of all fish. He is bold as a lion, and fights harder for his life than a salmon twice his size. A fish of three pounds will run out a considerable piece of line, and make a splendid leap, or series of leaps,—and then is the trying time. As often as not, your flies and the fish part company in the air, and you have to sit down muttering "curses not loud but deep," till an application to the flask soothes your wounded spirit, and invigorates you for fresh effort. A beautiful sight it is to see a sea-trout rise. No half-hearted attempt is his, but a determined rush for the fly, and down again like thought, leaving you the tiniest part of a moment to strike, and hardly time to admire his beautiful silvery coat. If you have been fortunate enough to get the steel into him, you will have time to admire him when you get him into the boat. Fishing for sea-trout with the fly is, we consider, the most exciting of all kinds of fishing—that is, if the fish run to a fair average weight. But we are sorry to say that lochs where it is to be enjoyed are, with the solitary exception of Loch Lomond, usually far out of ordinary reach,— and in the case of Loch Lomond, it is only *habitués* who usually come much speed on it; but once the angler gets a fair day there, he finds his way back often. True, there are some excellent sea-trout lochs in the north, and on the west coast and islands, but they are a far cry from civilisation. Nevertheless, if our readers can spare the time, let them find their way into some unfrequented spot where sea-trout are plentiful, and they will agree with us in thinking that that class of fishing is a most excellent sport. Some parts of Ireland are famous for their fine sea-trout fishing — white trout they call them there; and though we have never been there ourselves, we mean to go some day, when the Land Bill has pacified the natives, and made them law-abiding subjects. Meantime one runs the risk of being mistaken for a non-resident landlord, and that would be a pity for one's wife and family. But without any joking, this Irish sea-trout fishing is a pleasure to which we look forward; and in this work-a-day world, something to look forward to is half the enjoyment of life.

The capture of the SALMON is the ambition of all anglers, but we doubt very much if the sport is to compare with ordinary loch or sea-trout fishing, provided always that the latter are of good average weight. The tackle used in salmon-fishing is proportionately heavy, and after the first few rushes, if the fish be well hooked, there is little in it except a matter of time. Indeed it is said that some anglers, after hooking a salmon, hand the rod to a gillie to work and land the fish. This seems going too much in the other direction, but it is quite understandable. True, the size to which salmon run is a great inducement to go after them; but even in Loch Tay, where the biggest average is to be found, the sport, if such it can be called at all, is very questionable. The rod, line, gut, and minnows used are on such a strong scale, that a well-sized vessel might be moored with them without their breaking; and with several scores of yards of line ready for a rush, what earthly chance has the fish of escape, unless through the grossest carelessness? The fish may be loosely hooked, and get off, but this is quite a matter of chance, and the odds are that a hungry spring fish will not miss the lure. Thus the charm of salmon-fishing is in the raising and striking; and of all kinds of striking, the striking of the salmon is the most difficult: the fish being so large and silvery, the angler is certain to see him coming *at* the fly, and is very apt to strike too soon. But if it is borne in mind to strike *after* the broken water is visible, and not before it, this will soon be overcome. When you do strike, don't let it be a mere tightening of the line, as in trout-fishing, but a decided stroke. Some say that the salmon will hook himself by his own weight. This may be so, though we doubt it,—but don't trust to it. Certain it is, that the first rush of a fish does not usually fix him certain; and should the hook happen to be in a piece of hard gristle or on a bone, you will soon find this out for yourself, but generally at the cost of the fish.

Salmon-fishing is an expensive luxury; but if you can get it good, never mind the expense, but give it a trial. If you get good sport, you may not care to go in for smaller game again; but in all our experience we never knew a salmon fisher who did not enjoy trout-fishing as much in its own way as ever he did that of the nobler animal. There is something in the gossamer gut and small flies irresistibly attractive to all sportsmen, and from which no amount of salmon-fishing can ever wean them.

The *salmo ferox* is a fish on which many opinions have been expressed; and we have heard more than one old boatman say that he did not believe it to be anything but a big loch-trout, as, they ask, Who ever saw a young one? We see the young of all other fish, but why do we never come across a young *ferox*? It seems pertinent enough questioning, and we do not pretend to settle their doubts in either one way or another. Certain it is, he is a big strong fish with some features distinct from the ordinary loch trout, and that when caught he shows an amount of fight not to be equalled by any of his neighbours, either white or brown. He is usually caught by trolling either natural or artificial minnow; and the tackle should be mounted on gimp and fixed to a strong line, and plenty of it. We have read of a *ferox* rising to the fly, but never saw one so captured. There seems no reason why a gaudy fly should not attract him. After he is hooked the fun begins. A *ferox* of 10 to 12 lb. will give you amusement and excitement for an indefinite time; and you are never sure of him till he is in the boat. A friend of ours (a capital angler to boot) fishing with us on Loch Assynt in Sutherlandshire in 1877, hooked a fine specimen; and after battling with him for an hour, had the mortification of seeing fish, angel-minnow, and trace, disappear! A good boatman is a wonderful help in such a case; indeed without his help your chances are small. To be sure it is slow work trolling for *feroces*, and a whole day—yea, days—may be spent without getting a run. The angler must always be the best judge as to whether the chance is worth his while. Loch Awe, Loch Ericht, Loch Rannoch, and Loch Assynt, are good lochs for trying one's luck in this kind of fishing.

Then to come from the nobler to an inferior species, we get to PIKE fishing. Angling for this fish seems to be in great repute among our southern brethren, if we may judge by the literature on the subject; but somehow or other it is looked upon among our northern anglers with somewhat the same aversion that a Jew has to bacon, and fishing for pike is only resorted to when all chance of catching anything worthier is gone. We don't profess to say whence this antipathy arises; but we have heard stories from boatmen about the foul feeding of pike that makes the idea of eating him repulsive. Not but that we have eaten him, but we never did so with relish, however cunningly the *artiste* may have served him up. As a stock for soup he is good; but in Scotland it is better not to say what the origin of the stock is till your friends are at their *café noir*. But here we are only interested so far as the sport he gives is concerned; and unless the pike be all the larger—say not under 8 lb.—the sport is poor enough. Even a pike of 8 lb. and over, when hooked (which is done by trolling or casting a minnow and working it after the manner of a fly), makes one or two long pulls, not rushes like a fish of the *salmo* tribe; and after that he subsides into a sulk from which you must trust to the strength of your tackle to arouse him. The tackle should be mounted on gimp, for his teeth are very sharp;

and when removing the lure from his mouth, you will find it much safer to have previously put the foot-spar between his jaws to prevent him getting at your fingers.

There is a fly, if such it can be called, used in pike-fishing. This fly resembles a natural insect as much as a tea-pot resembles an elephant, but it does attract pike—in the same way, we suppose, that a piece of red flannel will attract a mackerel. If our readers wish to try it, they can buy it at almost any tackle shop. Pike are to be found in almost all lochs, though in the more frequented of our Scotch waters they are being slowly but surely exterminated. In others, again, they reign almost alone. But pike-fishing by itself is a poor affair, and we advise our readers only to take to it when they can do nought better. If any of them wish to go below the level of pike-fishing, we must refer them to the copious instructions of many books, from Isaak Walton downwards. For ourselves, when it comes to bait-fishing—except in running water, when worm-fishing is an art—we prefer catching whitings and haddocks in some of our beautiful salt-water lochs, to all the perch, roach, chub, and such-like, that ever swam. But in this please note that we are only expressing our own opinion, and with all respect to the opinions of many worthy anglers. We may say this, however, with all safety, that in angling, as in most other things, if one aims at the highest point of the art he is not at all likely to condescend to the lowest.

CHAPTER XI.

AFTER A DAY'S FISHING.

WHAT a pleasant fatigue succeeds a day's fishing! There is not, or should not be, a feeling of weariness, but just the satisfaction one feels after enjoying a health-giving recreation. Health-giving it certainly is to the body, and we have no hesitation in saying to the mind also. It makes one forget for the time being all the evils to which flesh is heir, and braces up the whole system to meet them when the necessity arises. But we must not go in for more sentiment than is actually needful. The practical duties after a day's fishing are these. If the weather has been damp, change all wet garments *at once*, and if at all practicable have a hot bath before sitting down to dinner. We say dinner advisedly, for the angler should always have a good sound dinner after a day's fishing, as however pleasant the work may have been, still it is exhausting to the body, and a rough tea, though good in itself, cannot pretend to have the reviving elements in it that a substantial dinner has. A glass of whisky, or even two, in cold water, will be found a very safe accompaniment. A good plan is to order your whisky by the bottle, and put your card in a nick made in the cork: the ordering of whisky in glasses is expensive and unsatisfactory. Your dinner over, turn your attention to your tackle. Unwind your lines, so far as they have been wet, from the reels, and lay them out on your bedroom floor; if any chance of being interfered with, wind them round the backs of chairs instead. They will be dry by the morning. Dry your reels thoroughly, and put in a little oil wherever you think they would be the better of it; and this should be done to any other article—spring-balance, gaff, &c.—that is liable to rust. Your creel or fishing-bag should be washed out and hung up to dry by the servants of the house immediately after the fish have been removed, which latter should be done without delay. Your landing-net should also be suspended in the open air, that it may get dry as speedily as possible. A landing-net will last double the time if attention is given to it in this way. Take out all used casting-lines from your book, and lay them on the mantelpiece till morning: this will insure the feathers being freed from moisture. And in the case of expensive flies, this is a matter of consideration, both on the point of expense as well as your possible inability to replace them where you may happen to be sojourning for the time. If you mean to make up a new cast or casts for the morrow, place the casting-lines in a little water in your basin. They will be in excellent order next morning for manipulation. Also soak in like manner the *gut* on which the flies which you mean to use are dressed. True, you may not be sure what flies you will put on till you see what sort of a day it may prove to be, but there is no harm done if you soak the gut (but only the gut) of as many flies as will give you a good choice.

We should have said nearer the beginning of this chapter to look well after your waterproofs, that they are not hung up in a hot place. A dry room or outhouse where there is a good draught is best. If your fishing should happen to be over for the time being, put your tackle past (after being thoroughly dried) in the most orderly fashion possible. For our own part, we have the drawer in our bookcase spaced out into compartments suitable for holding all our tackle, barring reels and such like; and this arrangement we find extremely useful, and wonderfully convenient when we wish to find anything. If, on the other hand, you are out on a lengthy holiday, and have time at your disposal, after putting things right for the day, and for next day too, we know of nothing better than a *good* rubber at whist for filling up the evening. It must be a *good* rubber, however, for the parlour game is neither relaxation nor pleasure. Hence we would advise all our angling friends to acquire a thorough knowledge of the game, as only to be learned with the aid of a good book on the subject. Remember that when staying at some out-of-the-way fishing hotel, you may be asked to form a table with good players, and not to be able to hold your own on such occasions is a great loss of pleasure to yourself, and usually a source of annoyance to the others. These remarks are somewhat apart from the subject of this book, but by way of an aside, they may be found not quite out of place.

Do not be beguiled into keeping late hours, for no one can fish well next day if he has not had a sufficient amount of sleep. But this is also an aside; for some men need more sleep than others, and each angler knows his own necessities best. We only promulgate the broad rule, that without proper rest no one can be in good trim with hand and eye for a pastime that needs both in a pre-eminent degree. We speak from experience in this too; and have sometimes imagined that our right hand had lost its cunning till we remembered that we had not been properly rested the night before.

CHAPTER XII.

REMINISCENCES.

HAVING exhausted, so far as we can imagine, the practical part of our little treatise, we proceed—in accordance with an idea which we had in our minds at starting — to give a few personal recollections, and to name one or two lochs where we have enjoyed good sport, and where it is still to be had for the trouble of going. Reminiscences are, as a rule, not specially interesting to the general reader, hence we shall not make them too lengthy; for we wish, above all things, that our readers shall close this volume without experiencing a shadow of weariness. One thing, however, we would like to say to our younger angling friends—Have as many personal adventures to look back to as you possibly can. The adventures themselves can be best sought after when the blood flows fast; for the time will come when the rod and the tackle will perforce have to be laid aside, and memory will then, unaided, afford you many a pleasant retrospect, and you will—even companionless—fight your battles over again. You remember the story of the illustrious Prince Talleyrand: when a young man acknowledged to him that he could not play whist, Talleyrand said to him—"Young man, what a sad old age you are preparing for yourself!" We don't mean to go this length as regards fishing; but we safely say that a man who lives to old age without having been a keen angler, has not only deprived himself of great enjoyments during his active life, but has neglected to lay up a provision for the time when the memory of them would have made life's closing seasons sweeter.

Our first acquaintance with LOCH ARD was very pleasant—not, perhaps, so much from any great expectation of sport, because at that time (many years ago now) we were young at the pastime, but more from the feeling of treading the ground made classical by the great Magician of the North, as the scene of the most stirring incidents in 'Rob Roy.' Attached to a big tree in front of the hotel at Aberfoyle there hangs a coulter, which tradition assigns as the veritable article which Bailie Nicol Jarvie made red-hot and used as a weapon of offence and defence when he was in a dilemma in what was, at that time, a very inaccessible part of the Highlands. Since then many a Glasgow magistrate has visited the spot—the inspection of the line of the noble waterworks undertaking which supplies the city being a sufficient excuse for the annual advent of the civic rulers. A railway station (Bucklyvie) is within eight miles of Aberfoyle, and Aberfoyle is within three miles of Loch Ard, and by the time this book is in the hand of the reader there will most likely be a railway station at Aberfoyle itself. Shade of Bailie Nicol Jarvie! what would you say if you were now to be allowed to haunt the old spot? to hear a locomotive screech where formerly you thought yourself so far "frae the Sautmarket o' Glesca"? We don't like the idea ourselves, and doubt very much if it will pay. However, it is the fishing alone which concerns us meantime, and we can at once assure our angling friends that the sport is good

—not but what one has to fish hard for a basket; but the same remark applies to all our near-at-hand lochs. On an ordinary good day a dozen to eighteen trout may be captured, and sometimes the baskets are heavier; but eighteen fish, weighing 9 lb. to 12 lb., is a very fair day's work. The trout average fairly a half pound, and pounders are by no means scarce: a two-pounder is come across occasionally, but he is the exception. The fish are very pretty, and for their size give excellent sport. Fine tackle is here absolutely essential to success, and as a matter of sport should always be used in fishing for common yellow trout. The loch, for its size, is much fished; and we fear that when the railway facilities are completed, there will require to be a considerable amount of re-stocking to keep it up to the old mark. The scenery is unsurpassed—wood, water, and mountain, making a picture of wondrous beauty. To the north of the loch, Ben Lomond rears its mighty summit; and in the spring-time (for Loch Ard is an early loch), before the summer sun has melted the winter's snow, the effect is grand in the extreme. April, May, and June, are supposed to be the best months for angling;

but we see no reason why, if the weather be favourable, these months should be singled out. The hotel accommodation at Aberfoyle is excellent. In the early months you must engage a boat beforehand: boatmen first-rate. Many a happy day we have spent on Loch Ard—sometimes successful and sometimes much the reverse; but in any case there is a witchery about the place that makes one enjoy himself in spite of all cares. Mind and body recruit their jaded energies, and get braced up to meet the stern realities of life.

In strong contradistinction, in this respect, to Loch Ard, is LOCH LEVEN. In the latter, if the angler is not catching fish, there is little of the beautiful to commend itself to the senses. The island on which the castle stands is pretty, and as a historic ruin is well worthy of a visit, but otherwise the scenery is very tame, and the surroundings not entrancing. But since we have drifted into speaking of Loch Leven, we may as well tell of the sport which is to be had there,—and this, as is well known, is exceptionally good. The quality of the fish is wonderful; and after reading the statistics of a year's fishing—last season something like 18,000 fish, weighing as many pounds, were killed—one is puzzled to know how it is kept up. The loch itself is a great natural feeding-pond, miles and miles of it being of an almost uniform depth, and a boat may drift almost anywhere, the angler feeling at the same time certain that fish are in his immediate vicinity. Trout of two and three pounds are quite common; and it is a rare occasion that a day's average does not come up to the pound for each fish. They are very fine eating, and cut red as a grilse. The company which rents the loch pay £800 to £1000 for the fishing, and they in turn keep a fleet of large boats—twenty we think—and let them out to anglers at the rate of 2s. 6d. an hour. Any number may fish from one boat. There are two boatmen in each boat,—one of whom is paid by the company, the other by the angler; and we are sorry to say that these men, with a few exceptions, are very much spoiled. There is a class of anglers (?) who frequent Loch Leven, whose whole aim seems to be, not sport so far as their own personal efforts are concerned, but the killing of as many fish as possible. If such a one has engaged a boat, he arms each boatman with a rod, and, of course, fishes himself, thus having three rods going at once. As we said before, the loch can be drifted without any attention from the men, after they have pulled up to the wind, and this enables them to get casting all the time that their employer is doing likewise. Not content with this, a couple of minnows are generally trolled astern when changing ground. We don't say that a man has not a right to do as he likes if he pays for his boat; but we *do* ask, Is this sport? And why should boatmen be spoiled in this way to such an extent that we have known them sulk a whole day because a spare rod was not allowed to be put up for their special benefit? But, of course, the men are just as they have been made, and true anglers, who fish for a day's sport, and not for the mere sake of slaughter, have the remedy in their own hands. Don't let anything deter you from fishing Loch Leven. It may be expensive; but if you get a good, or even a fair day, you will not regret the expense. Get a friend to join you, and the expense is not so heavy after all; and if your friend and yourself fish perseveringly all day, you will usually be rewarded with a very fine show of fish. There is no harm in letting your men fish when you are taking your lunch, *but don't allow a third rod to be put up.* The boatmen are, as a rule, only fifth-rate fishers, though, of course, a few of them handle a rod well. Our recollections of Loch Leven are pleasant in some ways, in others they are not; but don't fail to give it a trial, if only for the pleasure of handling a big fish on fine gut. The manager of the Loch Leven fishings, Captain Hall, fills a very difficult post with much acceptance to all concerned.

But to leave the Lowlands and go into the far North, we take you to LOCH ASSYNT, in Sutherlandshire, and to a little loch near it,— LOCH AWE by name. The journey to Assynt is long and weary: train to Lairg, and then between thirty and forty miles driving, is a good long scamper for fishing, but it is worth it. The inn at Inchnadamph is good, but when we were there in 1877 the boat accommodation was poor enough: perhaps they have improved upon that since. The first day after our arrival we had to go to Loch Awe, as the boats on the large loch (Assynt) were taken up. Such a morning of rain and wind! We were wet through our waterproofs during the four-mile drive, but luckily the weather moderated, and we had an excellent day's fishing. With two in the boat, we took 57 lb. weight of beautiful fish,—not large, but very game, and spotted intensely red. It must have been a good day, for many an angler tried his luck after our success, but never came near that mark, at least when we were there. Loch Assynt is more attractive, however, inasmuch as the chances of big fish are not remote. Trout of a pound weight, and over, are not uncommon, while the chance of a grilse adds excitement to the sport. Then *ferox*, as we have said in a previous chapter, are, comparatively speaking, not scarce, if one cares to go in for trolling for them. But, in any case, the angler is always sure of a basket of lovely yellow trout. On the hills behind the inn there is a small loch, called the MULACH-CÒRRIE, in which it is said that the gillaroo trout are to be found. Whether they are the real trout of that species or not, we cannot say, but certainly they are beautiful fish,—pink in the scales, and running to large sizes. We saw a basket taken by a friend, and it was a treat to look at. The fish were all taken with the fly, but we were told afterwards that worm is even deadlier than fly, and that one should never go there without a supply of "wrigglers." The hill between the inn and the Mulach-Corrie is a perfect paradise for fern-gatherers. It is said that about two dozen different kinds can be gathered; and we believe it, for even our untutored eyes discerned sixteen varieties! Our visit to Inchnadamph must be placed among the red-letter periods of our fishing life, and to be looked back to with much enjoyment.

LOCH MORAR, in Inverness-shire, is another delightful spot, and somewhat out of the usual track. The fishing is most excellent, and yellow trout of all sizes are very abundant. Sea-trout and salmon find their way frequently into the angler's basket; and half-way up the loch, which is a long one, at a bay into which the Meoble river flows, numbers of sea-fish are to be found. The best way is to fly-fish up to that bay one day, and seek shelter at night in some shepherd's cottage, thus being at hand to prosecute salmon and sea-trout fishing the next day, or days, if you find the sport good. It is right to take a supply of provisions and liquor with you, for the accommodation is humble. We write this from hearsay, as when we were there in mid-July salmon and sea-trout were not in the loch in large numbers; but still we caught some of the latter, and hooked, though, unfortunately, did not kill, any of the former. We should think that the beginning of August would be the best time for this loch as regards sea-fish; but the trout-fishing in July is unsurpassable. During our sojourn in 1876 at Arisaig, the nearest village to the loch, which is six miles off, and necessitating a drive over what was then a road sadly in need of General Wade's good offices, we had the services of a boatman, Angus by name, and his two boys, who could not speak a word of English,— Angus managing one boat, and his boys the other. We had the satisfaction—for indeed it was good fun—to be out with the boys one day; and the management of the boat had to be done by signals. It was wonderful how readily the boys got into the way of it, and how well we got on together. The memory of the hospitality which we enjoyed at Arisaig Inn will not be forgotten by any of our party; and we hope that the then occupier, Mr Routledge, will be there when we go back again. An inn was in course of being built at the loch-side in 1876, but we do not know how it has succeeded. The easiest way to Arisaig is by steamer, which usually goes once a-week; but the angler should, if possible, go to Banavie or Fort-William,— the latter for choice, as Banavie Hotel is famous for long bills (and we can testify that its notoriety in this respect is deserved),—and then drive to Arisaig. It is about thirty-eight miles from Fort-William to Arisaig, but the drive is something to be remembered during a lifetime. After having traversed this road, you will say, " There's no place like home" for grand and beautiful scenery. We must see Loch Morar again if we possibly can, before we bequeath our tackle to the next generation.

The time would fail us to tell of many other lochs, more or less famous for the good sport they afford; but the angler, if at all of an enterprising nature, need have little hesitation in taking up Mr Lyall's excellent 'Sportsman's Guide,' and making a selection on his own account. The information is very correct so far as we have tried it, sometimes—perhaps most anglers are inclined that way—erring a little to the *couleur de rose* side of things, but quite trustworthy in being followed as a suggester for a fortnight's fishing. We have gained much pleasure in exploring some of our more remote lochs, of the existence of which we might never have been aware but for its information. We cannot, however, close this long, but we hope not wearisome, chapter without singing the praises of our Queen of Scottish Lakes, LOCH LOMOND. The scenery of this beautiful spot is well known in some ways, but no amount of travelling in a steamer will reveal its beauties. To the tourist we would say, take a small boat at Luss and engage a man to row you among the islands which lie between Luss and Balmaha. With this hint to the tourist, we leave him, and turn the angler's attention to the sport—very precarious at most times, but excellent at others—to be had on Loch Lomond. Luss is the angling centre, and there are capital boats and men to be had by writing beforehand to the hotel-keeper, Mr M'Nab, who deserves much credit for the attention he pays to the wants of anglers.

The yellow-trout fishing is good, but, strange

to say, this class of sport is not much sought after. In April and May as good trout-fishing is to be had as on some other lochs that enjoy a greater reputation. But if the weather has been at all favourable to the fish running, the month of June sees the sea-trout fishing fairly commenced. It is a hard loch to fish; and if you are lucky enough to get two or three sea-trout in a day, consider yourself fortunate. They are a good average—2 lb. to 3 lb. being quite common—but they spread themselves so much over a large portion of water that one may fish a whole day and not come across them. This, however, is the exception, as in an ordinary fair fishing day in June, July, August, and September, and even October if the weather is mild, they are almost certain to be seen, if not caught. Some days really good sport is to be had—indeed, one is surprised at the show of fish; but fish or no fish, the charm of Loch Lomond is everlasting. The angler finds his way back over and over again, till, as in our own experience, the islands of Lonaig, Moan, Cruin, Fad, and last and least, Darroch, the great landing-spot, are as familiar to him as his daily business haunts. Then the chances of a salmon are good—indeed, this year (1881) a great many have been killed; but somehow or another the sea-trout fishing has not been so good, and though a salmon is always a salmon, we would rather see a good show of sea-trout at any time. Like our neighbours, we have had good and bad days on Loch Lomond; but disappointment has never soured us—indeed, the fascination seems to get stronger. And it is so very convenient for a day's fishing—down in the morning and home at night, with a good long day between. The charge for boatman is 5s. to 6s. and lunch; and though this seems high, it must not be forgotten that the distances are great. A boat costs 2s. per day. The men are good all over, some of them really first-rate. Many and many a story we could tell of happy fishing days, and of days most enjoyably spent when fishing was no go; but mostly every angler can do the same, and we don't wish to become too tiresome. Perhaps if we get the chance we may extend this chapter on some future occasion, and add some experiences of as yet untried places.

CHAPTER XIII.

CONCLUSION.

BROTHER of the gentle art, we bid you farewell! We have done our best to give you the benefit of our experience in the peaceful pursuit of loch-fishing; and if we have said too much or too little, pray excuse us, and in your goodness of heart reprove us for our verbosity, and tell us what is awanting. The spirit on our part has been very willing; but the memory may have been defective when it should have been most active, and quite possibly our love for the art may have somewhere or another led us into discursiveness where we should have been brief. We are all human, and he is a poor mortal who thinks he cannot err. Again we say farewell!—not for long, however, we hope. Who knows where we may meet? If we do, and you recognise us, don't forget to give us a little encouragement, and, if you can, new material for extending the usefulness of this publication. As we write, the hand of winter is upon us, and the rod and reel have been relegated to safe quarters; but spring will return, and the enforced cessation of our enjoyment will only add new zest to the music of the reel

"When green leaves come again."

PRINTED BY WILLIAM BLACKWOOD AND SONS.

Lately Published, Fifth Edition, Revised.

THE MOOR AND THE LOCH.

Containing Minute Instructions in all Highland Sports, with Wanderings over Crag and Corrie, Flood and Fell.

By JOHN COLQUHOUN.

2 vols. post 8vo, with Two Portraits and other Illustrations. 26s.

SOME OPINIONS OF THE PRESS.

"In the present delightful volumes, however, he presents all lovers of Scotland with the completest details of every Highland sport, on all of which he is an unexceptionable authority; and with what many will value even more, a series of life-like sketches of the rarer and more interesting animals of the country. He has thus brought up to the present level of knowledge the history of all the scarce birds and beasts of Scotland. . . . Henceforth it must necessarily find a place in the knapsack of every Northern tourist who is fond of our wild creatures, and is simply indispensable in every Scotch shooting-lodge."—*Academy.*

"We should recommend fishers to study carefully all the chapters on fishing for salmon, loch trout, sea trout, and yellow trout, whatever may be their experience or erudition. They will find general hints of immense use which they can apply to that local knowledge of their own river or 'water' which no books can teach, and which Mr Colquhoun himself would equally have to learn. But no chapter ought to be skipped, even by a reader who aspires to far less than the fourfold distinction of a Highland hunter, which consists in killing a red-deer, an eagle, a salmon, and a seal."—*Saturday Review.*

"The book is one written by a gentleman for gentlemen, healthy in tone, earnest in purpose, and as fresh, breezy, and life-giving as the mountain air of the hills amongst which the sport it chronicles is carried on."—*The World.*

"One of those rare and delightful books which, with all the fulness of knowledge, breathe the very freshness of the country, and either console you in your city confinement, or make you sigh to be away, according to the humour in which you happen to read it."—*Blackwood's Magazine.*

W. BLACKWOOD & SONS, Edinburgh and London.

LATELY PUBLISHED.

A HANDBOOK OF DEER-STALKING.

By ALEXANDER MACRAE,
Late Forester to Lord Henry Bentinck.

WITH INTRODUCTION BY HORATIO ROSS, ESQ.

Fcap. 8vo, with Two Photos. from Life. 3s. 6d.

"A work not only useful to sportsmen, but highly entertaining to the general reader."—*United Service Gazette.*

"The writer of this valuable little book speaks with authority, and sums up in a few pages hints on deer-stalking which the experience of a lifetime has enabled him to put forth. . . . We can only recommend every one who pursues the fascinating sport of which the author writes, to glance through, and indeed to read carefully, this handbook."—*Sporting and Dramatic News.*

"An interesting little book, alike because of the knowledge which its author displays of his subject, and of the simple style in which it is written. It is a handbook such as sportsmen must have long desired."—*Scotsman.*

RECREATIONS
OF
CHRISTOPHER NORTH.

With Portrait of the Author in his Sporting Jacket.

New Edition. Two Vols., crown 8vo, 8s.

"Welcome, right welcome, Christopher North; we cordially greet thee in thy new dress, thou genial and hearty old man, whose 'Ambrosian Nights' have so often in imagination transported us from solitude to the social circle, and whose vivid pictures of flood and fell, of loch and glen, have carried us in thought from the smoke, din, and pent-up opulence of London, to the rushing stream or tranquil tarn of those mountain-ranges."—*Times.*

W. BLACKWOOD & SONS, EDINBURGH AND LONDON.

www.ingramcontent.com/pod-product-compliance
Lightning Source LLC
Chambersburg PA
CBHW031604110426
42742CB00037B/1137